BOTANY FOR EVERYONE

Leaves

By Rachael Bush

Botany for Everyone: Leaves

ISBN: 978-1-960998-01-9

Written and illustrated by botanist Rachael Bush

Cadman, a dyslexic-friendly font created by Paul James Miller, is used in this book

From the Botany for Everyone Series, Volume 1

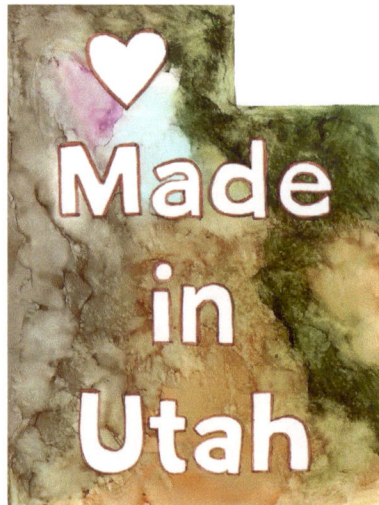

Made in Utah

Dedication and Acknowledgments

This book is dedicated to Ms. Joy Baty, one of the greatest teachers I've ever met. I hope this book will help you explain fall leaf coloration to your students for years to come.

I would like to acknowledge Dr. Sue Harley, the smartest scientist I know and my main resource for scientific accuracy. I would also like to thank Jason Baker for his botanical expertise (and my epic bio pic), Sarah Sandoval for the many hours of design help and botanical knowledge, Peter Fajardo for his creative input, and Kathleen Harris for her wonderful feedback. Thank you to my mechanical editing team: Sarah Sandoval, Terra Luft, Megan Condie, and Rebecca Ryan. Without their help, this book would not have been possible.

This is a plant:

Let's learn about leaves

A-Z Check out definitions and pronunciation of <u>underlined</u> words in the glossary on pages 18 and 19.

1

Why do plants have <u>leaves</u>?

<u>For photosynthesis</u>

Fun Fact: People used to think that plants grew by eating soil. It wasn't until the early 1600s when a Dutch scientist, Jan Baptist van Helmont, did an experiment that disproved this idea.

Photosynthesis:

Light Energy + Water + Carbon Dioxide $\longrightarrow$ Sugar + Oxygen

Most plants have leaves so they can make sugars through a process called <u>photosynthesis</u>. This process happens during the day.

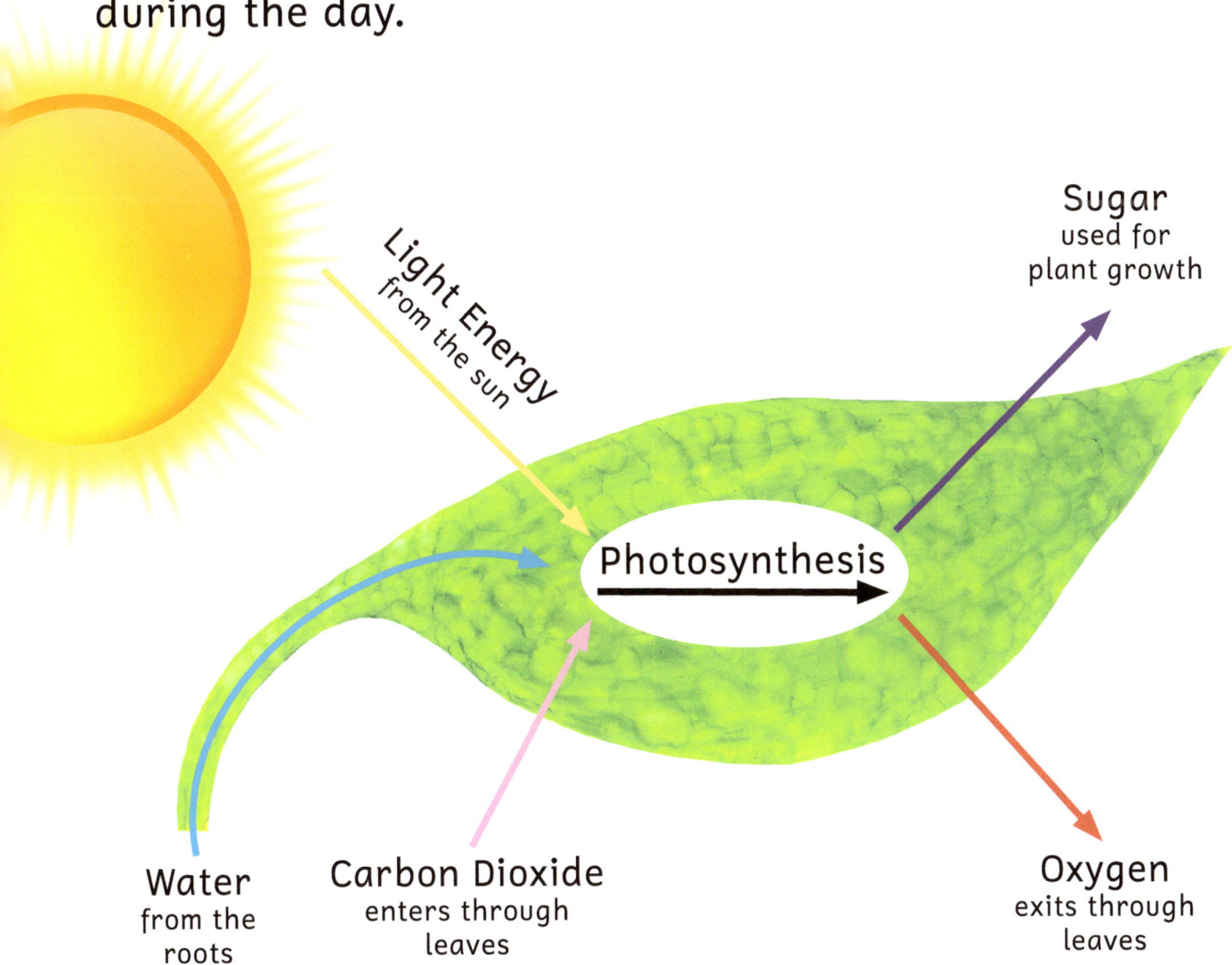

Light Energy
from the sun

Sugar
used for
plant growth

Photosynthesis

Water
from the
roots

Carbon Dioxide
enters through
leaves

Oxygen
exits through
leaves

Inside a leaf:

Stomata

Xylem

Ground
Tissue

Phloem

Vascular
Tissue

Epidermis

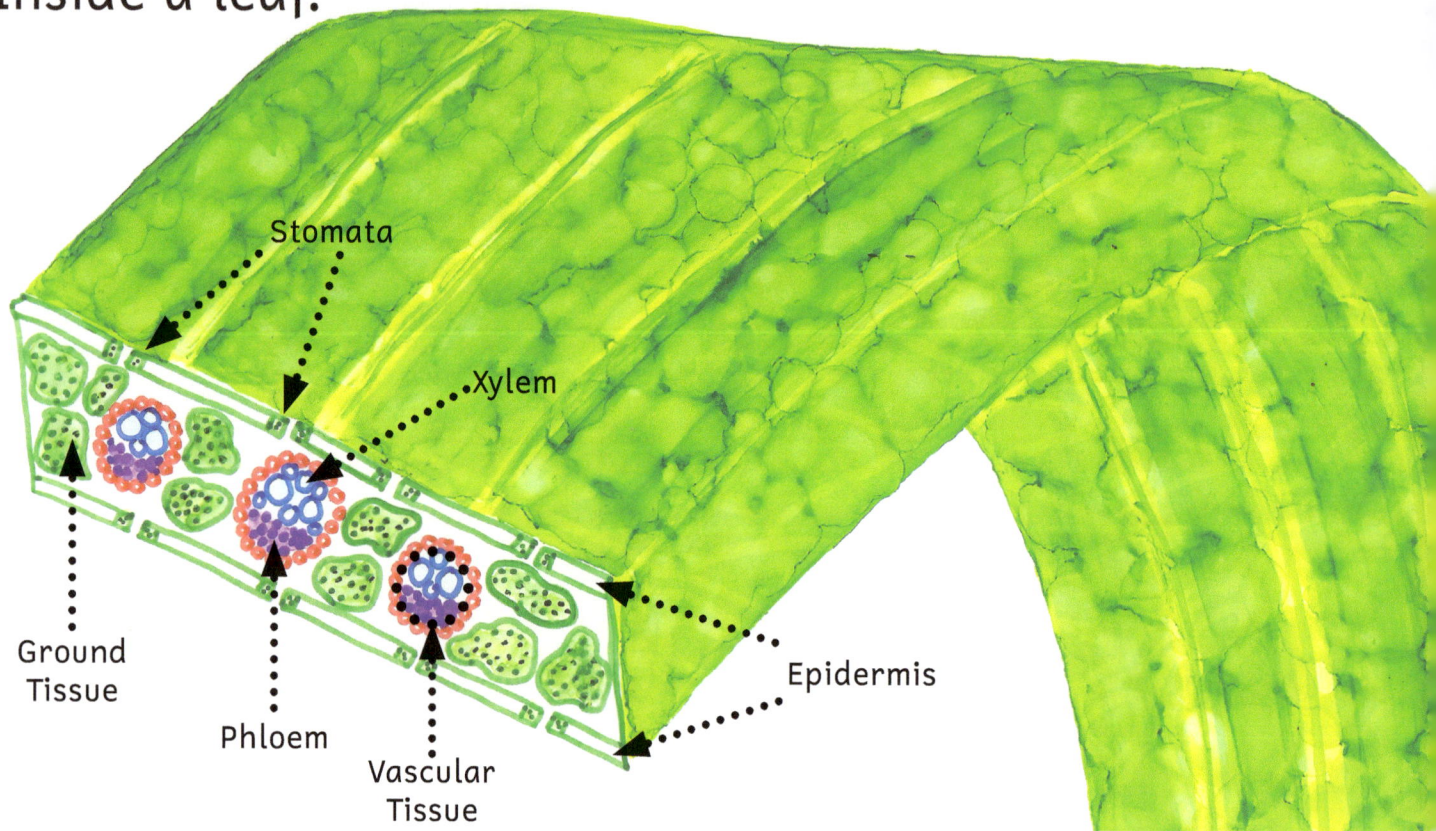

Epidermis: outer protective layer; in leaves and stems this is where stomata are found; in roots this is where root hairs originate

Ground Tissue: can do storage, support, and photosynthesis

Stomata: the pores in the epidermis of leaves and stems that allow carbon dioxide to enter and water and oxygen to exit the plant (the singular is stoma)

Vascular Tissue: composed of sugar-conducting phloem and water and mineral-conducting xylem; phloem and xylem run side by side in veins, transporting sugar, water, and minerals to all parts of the plant

4

Parts of a leaf:

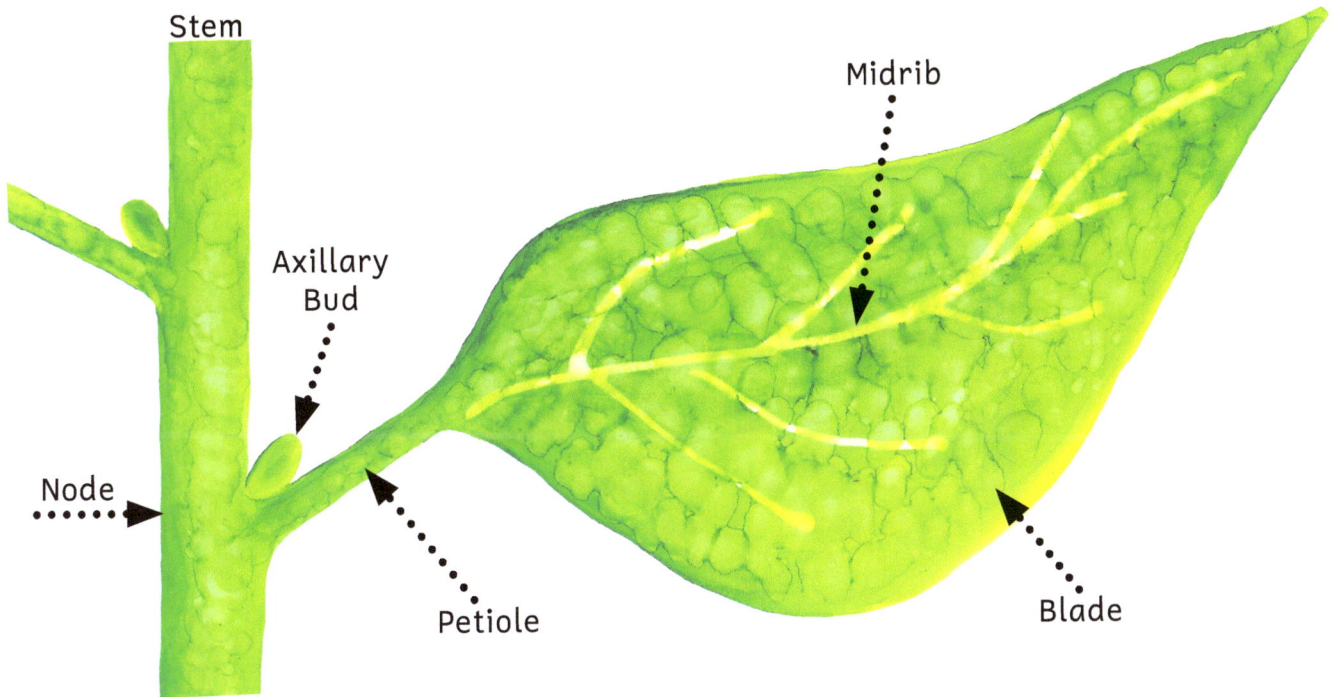

Axillary Bud: dormant bundle of <u>meristematic tissue</u> at the base of a leaf

Blade: the green, photosynthetic part of a leaf

Midrib: the main vein that transports water, minerals, and sugars to and from the stem and leaf

Node: the level (height) on the stem where one or more leaves and axillary buds attach

Petiole: the leaf stem, which may or may not be present

Stem: the organ of the plant that supports and displays leaves, flowers, and fruits; can store water and sugars

Not all leaves look the same. They have differently shaped <u>blades</u> and different <u>venation</u>.

Axillary bud position is very important when identifying leaves.

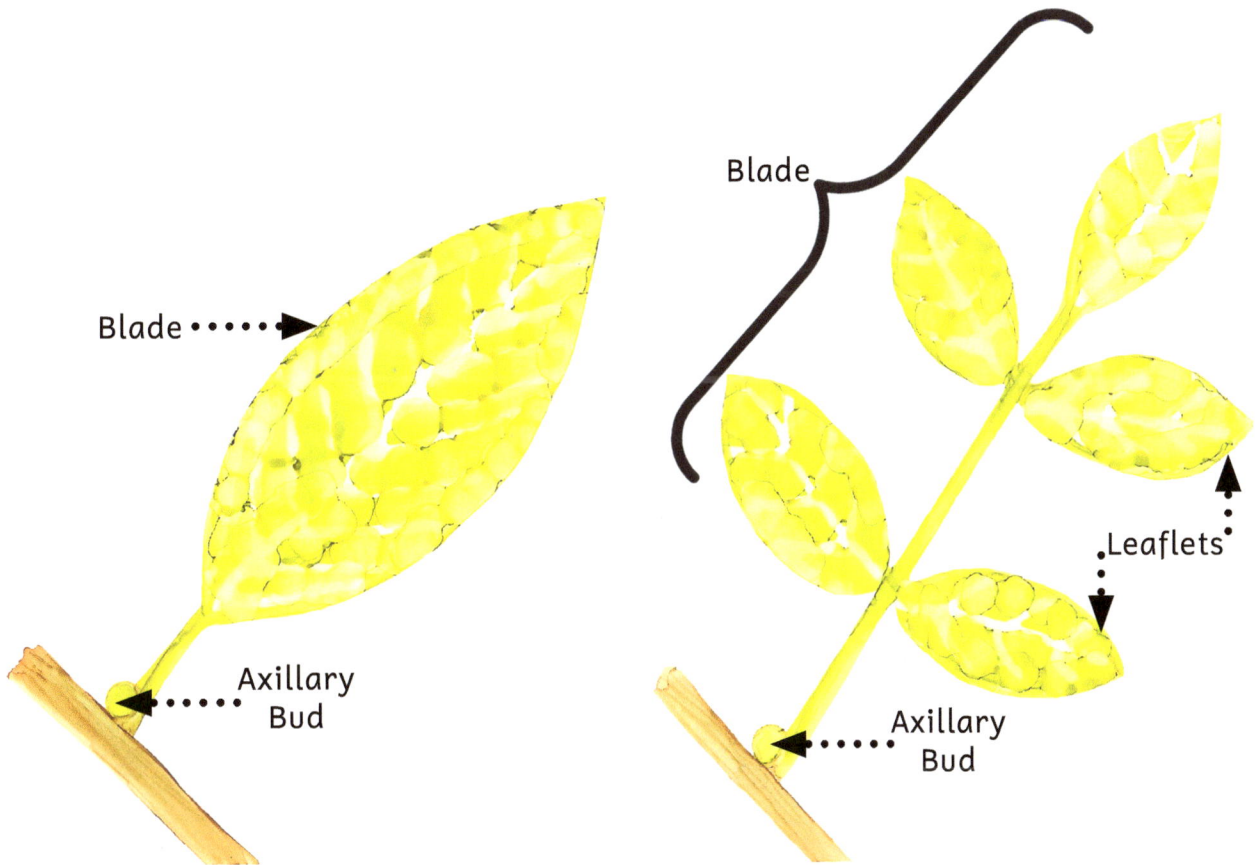

Blade ······▶

Axillary Bud

Blade

Leaflets

Axillary Bud

A <u>simple leaf</u> has an axillary bud at the base of a single blade (the petiole may or may not be present).

A <u>compound leaf</u> has an axillary bud at the base of a blade that is divided into many <u>leaflets</u> (the petiole may or may not be present).

7

Leaves attach to the stem in different ways.

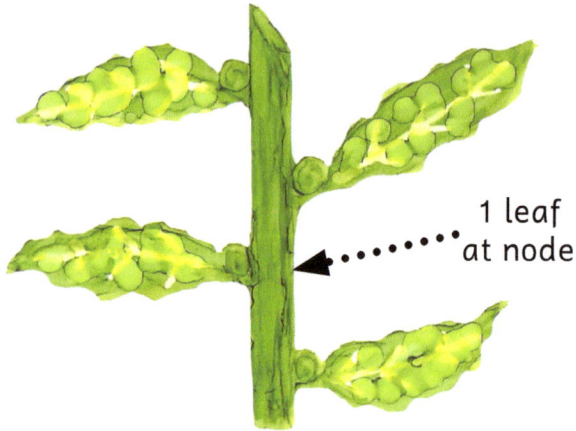

1 leaf
at node

2 leaves
at node

Alternate Leaf Arrangement: one leaf attaches to the stem at a node

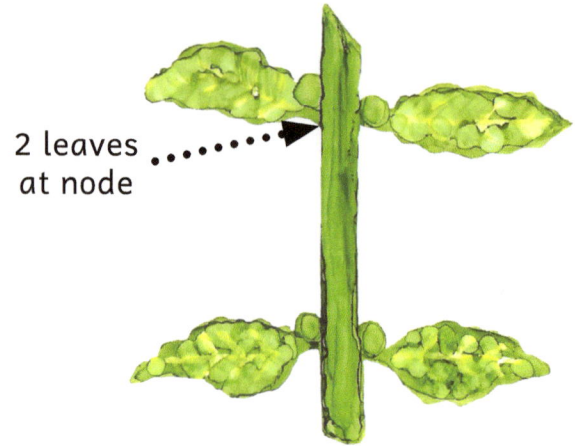

Opposite Leaf Arrangement: two leaves attach to the stem at a node

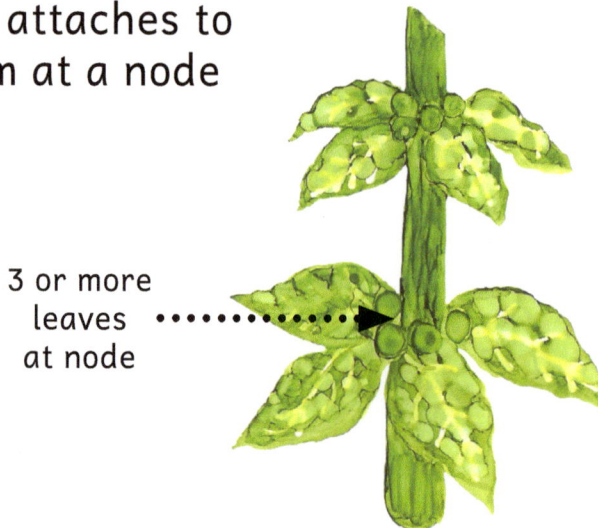

3 or more leaves at node

Whorled Leaf Arrangement: three or more leaves attach to the stem at a node

Some plants have modified leaves like these <u>tendrils</u> that hold onto nearby supports.

Tendrils

Fun Fact: Charles Darwin, the English naturalist who is most famous for his work in the Galapagos Islands in the 1800s, studied how climbing plants use tendrils to hold onto things.

Many cacti have modified leaves called <u>spines</u>. These leaves don't do photosynthesis. Instead, they protect the plant from herbivores.

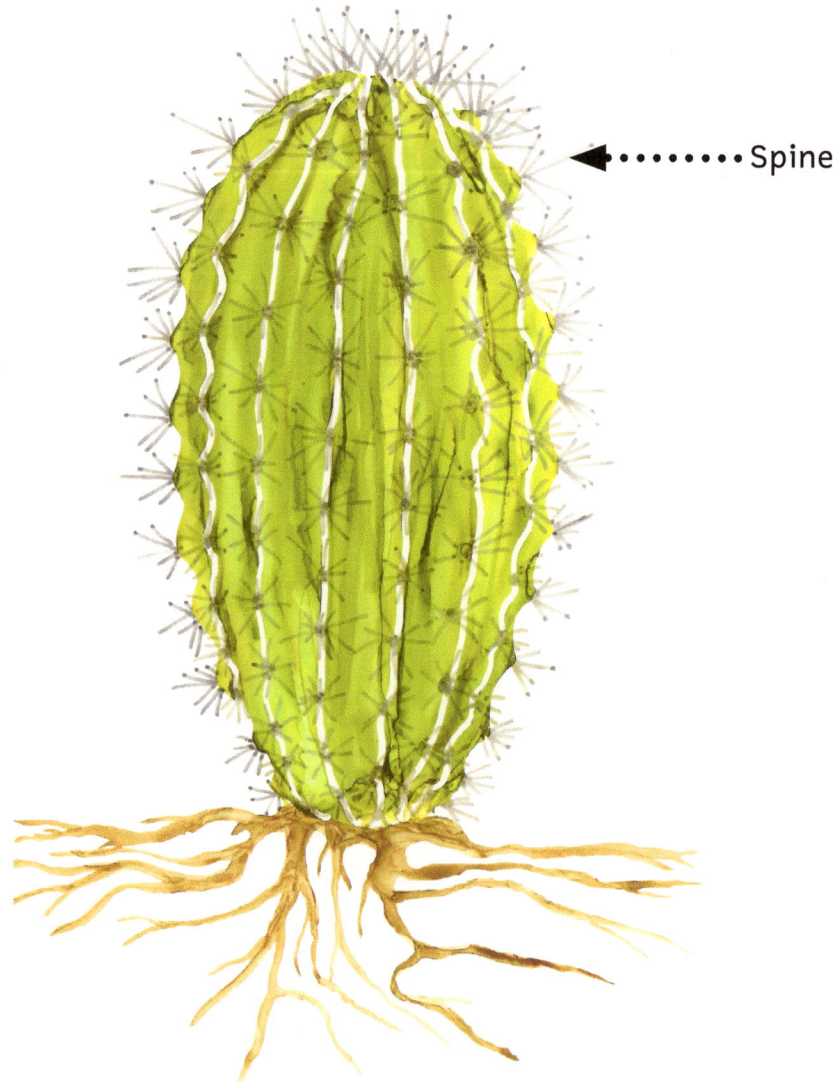
Spine

Fun Fact: Gila woodpeckers make their homes in saguaro cacti.

Some plants keep their leaves for years, others for only a few weeks. In plants that grow for many years, like trees, there are two ways they can lose their leaves:

Deciduous: plants that lose all leaves seasonally

Evergreen: plants that keep most leaves year round

Have you ever noticed when plants are losing their leaves, they change color? That's because there are many pigments inside leaves! Let's look at each one.

Chlorophyll is the green pigment responsible for photosynthesis. There is so much chlorophyll present, the other pigments are hidden. As leaves get ready to drop, chlorophyll is broken down faster than the other pigments in the leaves. Once the chlorophyll is gone, you can see the color of the other pigments.

Fun Fact: Have you ever had a green smoothie? They get their color from chlorophyll!

<u>Carotenoid</u> is a yellow or orange pigment that assists with photosynthesis. It is present all the time and is hidden by chlorophyll.

Fun Fact: Carotenoids are what make carrots orange.

<u>Anthocyanin</u> is a red, purple, or blue pigment that acts as a type of sunscreen to protect leaves. Some plants make anthocyanin seasonally. Others have this pigment all the time, making their leaves purple. A few plants don't make anthocyanin in their leaves at all.

Fun Fact: Anthocyanins make strawberries red, blackberries purple, and blueberries blue.

Tannin is a brown compound produced by plants to keep animals from eating the leaves. Tannins are bitter and hard to digest.

Fun Fact: People have used tannins for thousands of years to tan leather and make ink!

ACTIVITY WITH AN ADULT

Plant Press:

Once your supplies are ready, assemble your press like this, making up to 20 sections for leaves:

Plywood

Cardboard

Newspaper

Leaves

Newspaper

Cardboard

Newspaper

Leaves

Newspaper

Cardboard

Plywood

ACTIVITY WITH AN ADULT

Plant Press:

Supplies:

- ❀ 2 pieces of 1/8" thick plywood (botanists cut their plywood to 12"x18" but you can make it whatever size you want!)
- ❀ Cardboard (cut to the same size as your plywood)
- ❀ Newspaper (cut to the same size as your plywood)
- ❀ Two lashing straps to hold the layers together

Directions:

- ❀ Carefully position your leaves between pieces of newspaper, separating each section with a piece of cardboard.
- ❀ Cinch the straps around the press to squeeze it together.
- ❀ Wait for 2-7 days for the leaves to dry.
- ❀ Keep pressed leaves away from sunlight to preserve their colors

Glossary

Alternate Leaf Arrangement: one leaf attaches to the stem at a node

Anthocyanin (an(t)-thə-sī-ə-nən): a red, purple, or blue pigment that acts as a type of sunscreen to protect the leaf; also colors flowers and fruits

Axillary Bud (ak-sə-ler-ē): a dormant bundle of meristematic tissue at the base of a leaf

Blade: the green photosynthetic part of a leaf

Carotenoid (kə-rä-tə-noid): a yellow or orange pigment that assists with photosynthesis; also colors flowers and fruits

Chlorophyll (klor-ə-fil): the green pigment needed for photosynthesis

Compound Leaf: has an axillary bud at the base of the blade, which is divided into many leaflets; petiole may or may not be present

Deciduous (di-si-jə-wəs): plants that lose all leaves seasonally

Epidermis (e-pə-dər-məs): the outer protective layer; in leaves and stems this is where stomata are found; in roots this is where root hairs originate

Evergreen: plants that keep most of their leaves year round

Ground Tissue: can do storage, support, and/or photosynthesis depending on where it is found

Leaf: the photosynthetic organ of most plants (the plural is leaves)

Leaflet: part of a compound leaf blade

Meristematic Tissue (mer-ə-stə-ma-tik): actively dividing cells

Midrib: main vein that transports water, minerals, and sugars to and from the leaf

18

Glossary

Node: the level (height) on stem where one or more leaves and buds attach

Opposite Leaf Arrangement: two leaves attach to the stem at a node

Petiole (pe-tē-ōl): the leaf stem, which may or may not be present

Photosynthesis (fō-tō-sin(t)-thə-səs): the process in which plants use light energy, carbon dioxide, and water to make sugars and oxygen

Simple Leaf: has an axillary bud at the base of a single blade; petiole may or may not be present

Spine: a modified leaf that protects the plant from some herbivores; at maturity, they're not photosynthetic and are usually hard and pointy

Stem: the organ of the plant that supports and displays leaves, flowers, and fruits; can store water and sugars

Stomata (stō-mə-tə): the pores in the epidermis of leaves and stems that allow carbon dioxide to enter and water and oxygen to exit the plant (the singular is stoma)

Tannin (ta-nən): a brown compound produced by plants to keep animals from eating them

Tendril: a modified leaf that holds onto a support

Vascular Tissue: composed of sugar-conducting phloem (flō-em) and water and mineral-conducting xylem (zī-ləm); phloem and xylem run side by side in veins, transporting sugar, water, and minerals to all parts of the plant

Venation (ve-nā-shən): the arrangement of veins (vascular tissue) in a leaf

Whorled Leaf Arrangement: three or more leaves attach to the stem at a node

About the author:

Rachael Bush has a degree in botany with a minor in chemistry. She began her career managing the botany greenhouses and labs at Weber State University. Rachael then shifted her focus to teach children and young adults about plants and their importance in our world.

Rachael is also an alcohol ink artist who makes science-based art under the business name Molecular Inks. All of the images featured in this book were inked by her. You can find her art at MolecularInks.Etsy.com

Look for other books in the
Botany for Everyone series at
botanyforeveryone.com

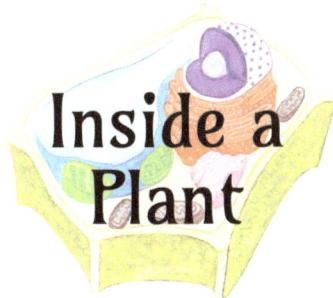

BOTANY FOR
EVERYONE

Stems

By Rachael Bush

BOTANY FOR
EVERYONE

Wood
& Bark

By Rachael Bush

BOTANY FOR
EVERYONE

Inside a
Plant

By Rachael Bush

BOTANY FOR
EVERYONE

Plant
Adaptations

By Rachael Bush

www.ingramcontent.com/pod-product-compliance
Lightning Source LLC
LaVergne TN
LVHW072133070426
835513LV00002B/88